HOW TO SOLDER ELECTRONICS

A COMPREHENSIVE GUIDE TO
SOLDERING

Copyright@2023

Harvey Talon

TABLE OF CONTENT

CHAPTER I. INTRODUCTION

IMPORTANCE OF SOLDERING IN ELECTRONICS

Soldering plays a crucial role in the field of electronics and is of utmost importance for several reasons:

1. Electrical Connection: Soldering allows for the creation of reliable electrical connections between components. It ensures a low-resistance pathway for the flow of electric current, enabling the proper functioning of electronic circuits.

2. Mechanical Stability: Soldered connections provide mechanical stability to electronic components, preventing them from coming loose or disconnecting due to vibrations, movements, or external forces. This stability is particularly important in applications where the electronic device may be subject to physical stress or environmental conditions.

3. Signal Integrity: In high-frequency circuits, such as those found in telecommunications, audio/video equipment, and data transmission systems, soldered connections are crucial for maintaining signal integrity. Solder joints offer a consistent and low-resistance path for the transmission of signals, minimizing losses and interference.

4. Heat Dissipation: Soldering allows for efficient heat dissipation by creating a direct thermal path between components and heat sinks. This is essential in power electronics and high-power applications where excessive heat can damage components or degrade performance.

5. Miniaturization: With the trend of electronics becoming smaller and more compact, soldering enables the assembly of tiny surface mount components (SMDs) onto circuit boards. The precision and control offered by soldering techniques are vital for handling these miniaturized components and achieving a high-density interconnection.

6. Repair and Rework: Soldering is essential for repairing and reworking electronic devices. When components fail or need to be replaced, soldering allows for the removal and replacement of faulty parts without damaging the surrounding circuitry.

7. Customization and Prototyping: Soldering skills empower electronics enthusiasts, hobbyists, and innovators to customize and prototype their own electronic projects. By soldering components together, they can create unique circuits, experiment with designs, and bring their ideas to life.

8. Industry Standards: Soldering is an integral part of manufacturing processes in the electronics industry. Various industry standards and certifications ensure the reliability, quality, and safety of soldered connections in consumer electronics, automotive electronics, aerospace systems, medical devices, and more.

BENEFITS OF MASTERING SOLDERING TECHNIQUES

Mastering soldering techniques offers several benefits for individuals involved in electronics:

1. Enhanced Professional Skills: Soldering is a fundamental skill required in many electronics-related professions, such as electronic engineering, circuit board design, repair and maintenance, manufacturing, and prototyping. By mastering soldering techniques, professionals can significantly enhance their skill set, making them more valuable and competent in their respective fields.

2. Improved Efficiency: Efficient soldering techniques enable faster and more accurate assembly and repair of electronic circuits. With precise soldering skills, professionals can work more efficiently, reducing the time required to complete tasks and increasing overall productivity.

3. Reliable and Durable Connections: Properly soldered connections are more reliable and durable than alternative methods, such as wire wrapping or using connectors. Mastering soldering techniques ensures robust connections that can withstand environmental factors, mechanical stress, and temperature fluctuations, resulting in longer-lasting and more dependable electronic devices.

4. Better Signal Integrity: Soldered connections offer excellent electrical conductivity, minimizing resistance and signal losses. By mastering soldering techniques, professionals can create high-quality connections, ensuring optimal signal integrity in electronic circuits. This is particularly important in applications where accurate transmission and reception of signals are critical, such as telecommunications, audio/video equipment, and data communication systems.

5. Versatility and Adaptability: Soldering skills provide individuals with the ability to work with a wide range of electronic components, including through-hole and surface mount devices. Mastering soldering techniques allows professionals to adapt to different project requirements and handle various component packages, opening up more opportunities for diverse electronic projects.

6. Cost Savings: With proficient soldering skills, individuals can perform their own repairs and modifications, saving money on professional repair services or component replacements. Additionally, soldering expertise enables individuals to salvage and rework electronic components, reducing waste and increasing cost-effectiveness.

7. Creative Freedom: Mastering soldering techniques grants individuals the freedom to create and innovate. It empowers them to design and build custom electronic circuits, prototypes, and DIY projects according to their unique requirements and ideas. Soldering skills enable individuals to bring their creative visions to life and explore new possibilities in electronics.

8. Personal Satisfaction: Successfully soldering electronic components and witnessing a functional circuit or project can provide a great sense of accomplishment and personal satisfaction. Mastering soldering techniques allows individuals to experience the joy and fulfillment that comes with creating, repairing, and working with electronic devices.

CHAPTER II. UNDERSTANDING SOLDERING BASICS

WHAT IS SOLDERING?

Soldering is a process of joining two or more metal components together using a filler material called solder. It involves heating the metal parts to be joined and applying the solder, which melts and flows between the components, creating a permanent bond when cooled. The primary purpose of soldering in electronics is to create reliable electrical connections between components on circuit boards or other electronic devices. Soldering not only establishes the electrical connection but also provides mechanical stability to the joint, ensuring that the components remain securely attached. Solder is typically made from a combination of metals, such as tin and lead or tin and silver, although lead-free solder has become more prevalent due to environmental concerns. The solder has a lower melting point than the metal being joined, allowing it to melt and flow when heated while maintaining a solid state at room temperature. To perform soldering, a soldering iron or soldering station is used. The soldering iron consists of a heated metal tip that transfers heat to the joint, melting the solder. The soldering iron is held in one hand, while the other hand holds the solder wire. The heated tip of the soldering iron is brought into contact with the joint, and the solder wire is applied to the joint, allowing the solder to flow and create the connection. Flux, a chemical cleaning agent, is often included in the solder wire or applied separately to remove oxidation from the metal surfaces and promote good solder flow. Soldering can be performed on various types of electronic components, including through-hole components and surface mount devices. Through-hole soldering involves passing the component leads through holes in a circuit board and soldering them on the other side. Surface mount soldering involves soldering components directly onto the surface of the circuit board, without the need for holes. Mastering soldering techniques is essential for professionals, hobbyists, and anyone working with electronic devices. It enables them to create reliable electrical connections, repair circuits, and build electronic projects with precision and efficiency.

DIFFERENT TYPES OF SOLDER

There are several different types of solder available, each with its own characteristics and applications. The choice of solder depends on factors such as the type of metals being joined, the desired strength and reliability of the joint, and any specific environmental or regulatory requirements.

1. Tin-Lead Solder (Sn-Pb):

- Traditional solder composition.
- Provides good wetting and flow characteristics.
- Offers eutectic (63% tin and 37% lead) and near-eutectic options.
- Easier to work with due to lower melting temperature.
- Not compliant with RoHS (Restriction of Hazardous Substances) regulations due to lead content.

2. Lead-Free Solder:

- Developed as a safer alternative to tin-lead solder.

- Compliant with RoHS regulations.
- Various formulations available, such as tin-silver-copper (Sn-Ag-Cu) or tin-copper (Sn-Cu).
- Higher melting temperatures compared to tin-lead solder.
- May require additional care and attention during soldering due to different properties.

3. Tin-Silver Solder (Sn-Ag):

- Contains tin and silver as the primary metals.
- Offers improved strength and conductivity compared to tin-lead solder.
- Widely used in high-reliability applications, such as aerospace and medical electronics.
- Suitable for applications where temperature cycling and mechanical stress resistance are important.

4. Tin-Copper Solder (Sn-Cu):

- Contains tin and copper as the primary metals.
- Provides good conductivity and mechanical strength.
- Suitable for general-purpose soldering applications.
- Often used in plumbing and HVAC (heating, ventilation, and air conditioning) applications.

5. Flux-Core Solder:

- Solder wire with a flux core, which eliminates the need for separate flux application.
- Flux helps remove oxidation and ensures proper wetting and flow of solder.
- Available in different solder compositions, including tin-lead and lead-free options.
- Convenient for general-purpose soldering, especially for quick repairs and prototyping.

TOOLS AND EQUIPMENT REQUIRED FOR SOLDERING

To perform soldering, several tools and equipment are necessary to ensure safe and effective soldering operations. These are the essential tools and equipment required for soldering electronics:

1. Soldering Iron:

- The primary tool used for heating and melting solder.
- Available in various wattages and tip sizes, depending on the application.
- Temperature-controlled soldering stations are preferred for precise temperature regulation.

2. Soldering Iron Stand:

- A stand or holder to hold the soldering iron when not in use.
- Prevents accidental burns or damage to the work surface.

3. Solder:

- The filler material used to create the solder joint.
- Comes in various compositions, such as tin-lead or lead-free solder.
- Available in different diameters or forms, such as solder wire or solder paste.

4. Flux:

- A chemical cleaning agent used to remove oxidation and ensure good solder flow.
- Flux promotes wetting of the solder to the metal surfaces being joined.
- Flux can be in the form of a liquid flux, flux pen, or flux core in solder wire.

5. Soldering Tip:

- The replaceable tip of the soldering iron that directly contacts the components and joints.
- Available in various shapes and sizes to suit different soldering tasks.
- Tips can be conical, chisel, or specialized for specific applications.

6. Desoldering Tools:

- Tools for removing solder or desoldering components:
- Desoldering pump or solder sucker: Creates a vacuum to suck up molten solder.
- Desoldering braid or solder wick: Absorbs molten solder through capillary action.

7. Helping Hands: A tool with adjustable alligator clips or clamps to hold the work piece or components in place during soldering.

8. Wire Cutters and Strippers: Used for cutting and stripping wire insulation, preparing components for soldering.

9. Tweezers: Fine-tipped tweezers for holding small components or precise placement during soldering.

10. Multimeter: Used to measure voltage, resistance, and continuity in circuits. Helps in troubleshooting and testing the integrity of soldered connections.

11. Safety Equipment: Safety glasses or goggles to protect eyes from soldering fumes and potential splatters. Heat-resistant gloves to protect hands from burns and to handle hot components.

12. Ventilation: Adequate ventilation or a fume extractor to remove soldering fumes from the work area.

SAFETY PRECAUTIONS AND BEST PRACTICES

When engaging in soldering operations, it's crucial to follow safety precautions and best practices to protect yourself, prevent accidents, and ensure successful soldering outcomes. There are some important safety guidelines and best practices to consider:

1. Work in a Well-Ventilated Area:

- Perform soldering in a well-ventilated space or near a fume extractor.
- Soldering fumes may contain harmful substances, such as flux vapors, so proper ventilation is essential.
- If ventilation is inadequate, consider using a fume extractor or wearing a respirator mask suitable for soldering operations.

2. Wear Personal Protective Equipment (PPE):

- Use safety glasses or goggles to protect your eyes from potential splatters, solder, or debris.
- Consider wearing heat-resistant gloves to protect your hands from burns and handle hot components.
- Avoid loose clothing or jewelry that may come into contact with the soldering iron.

3. Maintain a Clean and Organized Workstation:

- Keep your work area clean and free from clutter to minimize the risk of accidents or damage.
- Have a designated space for tools and equipment to prevent accidental contact with hot surfaces.
- Properly store solder and other chemicals in a safe and organized manner.

4. Handle the Soldering Iron with Care:

- Always assume the soldering iron is hot and handle it carefully to avoid burns or electrical shock.
- Use the soldering iron stand or holder to secure the iron when not in use.
- Never touch the tip of the soldering iron with your bare hands or place it on flammable surfaces.

5. Be Mindful of Heat:

- Avoid placing flammable materials, such as papers or plastics, near the soldering area.
- Keep a fire extinguisher or fire blanket accessible in case of emergencies.
- Allow the soldering iron to cool down before storing it or performing maintenance.

6. Use Proper Soldering Techniques:

- Follow proper soldering techniques to create reliable and effective solder joints.
- Ensure the components and soldering iron tip are clean for good thermal transfer and wetting.
- Apply the right amount of heat and solder to achieve proper solder flow and joint formation.
- Avoid excessive soldering time on sensitive components to prevent damage.

7. Double-Check Polarity and Connections:

- Verify the correct polarity and orientation of components before soldering.
- Check for proper connections and ensure there are no short circuits or cold solder joints.

8. Practice Electrical Safety:

- Disconnect power sources and unplug devices before soldering to prevent electrical shocks.
- Handle electronic components with caution to avoid damaging them or causing injury.

9. Continuous Learning and Skill Development:

- Stay updated on best practices, safety guidelines, and new soldering techniques.
- Engage in continuous learning to enhance your soldering skills and knowledge.

CHAPTER III. PREPARING FOR SOLDERING

SETTING UP A SOLDERING WORKSTATION

Setting up a proper soldering workstation is important for creating a safe and efficient environment for soldering operations. There are some steps to set up a soldering workstation:

1. Choose a Well-Ventilated Area:

- Select a well-ventilated space for your soldering workstation.
- Adequate ventilation helps in dissipating soldering fumes and maintaining air quality.
- If possible, work near a window or use a fume extractor to remove fumes.

2. Clean and Organize the Work Area:

- Clear the workspace of any clutter, ensuring a clean and organized area.

- Remove any flammable materials or substances that may be hazardous during soldering.

3. Soldering Workbench:

- Set up a sturdy workbench or table as your soldering station.
- Ensure the work surface is heat-resistant and non-flammable.
- Cover the surface with a heat-resistant mat or soldering pad for added protection.

4. Adequate Lighting:

- Ensure sufficient lighting at your soldering workstation for clear visibility.
- Use adjustable desk lamps or dedicated task lighting to illuminate your work area effectively.

5. Soldering Equipment and Tools:

- Gather all the necessary soldering equipment and tools required for your projects.

- Arrange them within reach, so they are easily accessible during soldering.

6. Soldering Iron and Stand:

- Place your soldering iron stand or holder on the workbench.
- Position it securely to hold the soldering iron when not in use, preventing accidental burns or damage.

7. Storage for Tools and Supplies:

- Allocate storage space for your soldering tools, such as wire cutters, tweezers, and solder spools.
- Use tool organizers, drawers, or containers to keep your tools and supplies organized and easily accessible.

8. Safety Measures:

- Keep a fire extinguisher or fire blanket nearby in case of emergencies.

- Wear appropriate personal protective equipment (PPE), such as safety glasses or goggles, and heat-resistant gloves.

9. Electrical Outlets:

Ensure there are sufficient electrical outlets near your soldering workstation to power your soldering iron and any other tools or equipment you may use.

10. Fire-Safe Container:

- Have a fire-safe container or tray to place hot soldering irons or tips after use.
- This prevents accidental burns or damage to the workbench or surrounding materials.

11. Reference Materials:

Keep reference materials, such as soldering guides or datasheets, readily available for quick access to information and soldering tips.

When selecting a soldering iron and tips for your soldering projects, it's important to consider factors such as the type of work you'll be doing, the size of components, and your budget.

1. Power and Temperature Control:

- Determine the power requirements based on the size and complexity of your soldering tasks.

- Higher wattage soldering irons heat up faster and are suitable for larger components and heavier soldering jobs.

- Temperature control is crucial for precise soldering. Look for soldering irons with adjustable temperature settings for better control over the soldering process.

2. Tip Compatibility and Varieties:

- Check if the soldering iron supports interchangeable tips.

- Different tips are available in various shapes (conical, chisel, bevel, etc.) There are various sizes available to suit different soldering tasks.
- Having a range of tips allows versatility in soldering various component sizes and shapes.

3. Quality and Durability:

- Choose a soldering iron from reputable brands known for their quality and durability.
- Quality soldering irons typically have better temperature stability, heat recovery, and longer lifespans.
- Consider reading reviews or seeking recommendations from experienced soldering enthusiasts or professionals.

4. Ergonomics and Comfort:

- Consider the ergonomics of the soldering iron, especially if you plan to use it for extended periods.
- Look for a comfortable grip and a lightweight design that reduces strain on your hand during soldering.

5. Specialty Soldering:

- If you anticipate working on specialized soldering tasks, such as surface mount technology (SMT) or intricate electronics, consider soldering irons designed specifically for those applications.
- Specialty soldering irons may have features like fine tips or adjustable temperature profiles tailored for specific tasks.

6. Budget:

- Determine your budget and balance it with your soldering requirements.

- It's possible to find soldering irons that offer a good balance of quality and affordability within a reasonable price range.

7. Replacement and Availability:

- Check the availability and cost of replacement tips for the soldering iron model you choose.
- Availability of spare parts and accessories can extend the lifespan of your soldering iron and ensure its continued usability.

PREPARING ELECTRONIC COMPONENTS FOR SOLDERING

Preparing electronic components properly before soldering is crucial for achieving reliable and durable solder joints. There are the steps to prepare electronic components for soldering:

1. Collect the required tools and equipment:

- Wire cutters or strippers for preparing wires and leads.

- Tweezers or small pliers for handling small components.
- Use isopropyl alcohol and a lint-free cloth to clean the surface.

2. Inspect the components:

- Carefully examine the components for any visible defects, damage, or bent leads.
- Verify that you have the correct components and check for any discrepancies in part numbers or specifications.

3. Clean the components:

- Use isopropyl alcohol and a lint-free cloth or cotton swab to clean the leads or terminals of the components.
- Remove any dirt, oil, or oxidation that could hinder the soldering process.
- Ensure that the components are completely dry before proceeding.

4. Trim and shape component leads:

- Use wire cutters or strippers to trim excess lead length if needed.
- Shape the leads of components to fit the intended soldering locations, ensuring proper alignment and clearance.

5. Tin component leads (optional):

- Tinning the leads involves applying a thin layer of solder to the leads before soldering to facilitate the soldering process.
- Apply a small amount of flux to the leads.
- Heat the lead with a soldering iron and touch the solder wire to the lead, allowing the solder to flow and coat the lead evenly.
- Avoid excessive solder buildup that could cause bridging or short circuits.

6. Component positioning and alignment:

- Position the components on the circuit board or soldering area according to the schematic or layout diagram.

- Ensure the leads or terminals are aligned with the corresponding pads or holes.

7. Use a helping hand or clamps:

- Use a helping hand tool or small clamps to hold the components in place during soldering.

- This helps to keep the components steady and aligned, especially when working with small or delicate components.

8. Heat management:

- Consider using heat sinks or heat-absorbing clips on sensitive components that could be damaged by excessive heat during soldering.

- Heat sinks help dissipate heat and protect sensitive components from thermal stress.

IDENTIFYING COMMON SOLDERING PROBLEMS AND THEIR SOLUTIONS

During soldering, various issues can arise that may affect the quality of the solder joint or the overall functionality of the electronic device.

1. Cold Solder Joint:

Symptoms: Dull, grainy, or lumpy appearance of the solder joint.

Causes: Insufficient heat applied or premature movement of the component or solder before the solder solidifies.

Solutions: Reheat the joint, ensuring sufficient heat transfer to create a smooth, shiny solder joint. Avoid disturbing the joint until the solder solidifies.

2. Solder Bridging:

Symptoms: Excess solder connecting two adjacent pads or leads, resulting in a short circuit.

Causes: Excessive solder applied, inadequate clearance between leads or pads, or incorrect soldering technique.

Solutions: Use desoldering tools (such as a desoldering pump or wick) to remove excess solder and create a clear separation between the solder joints. Ensure proper spacing and use flux to prevent bridging.

3. Insufficient Solder Wetting:

Symptoms: Solder fails to flow smoothly and make proper contact with the component lead or pad.

Causes: Dirty or oxidized surfaces, inadequate flux, or insufficient heat.

Solutions: Clean the surfaces with isopropyl alcohol and a lint-free cloth. Apply flux to improve wetting, and reheat the joint with adequate heat to ensure proper solder flow and wetting.

4. Solder Balling:

Symptoms: Small spherical blobs of solder that form on the component or pad.

Causes: Excessive solder applied, excessive heat, or improper soldering technique.

Solutions: Use desoldering tools or a soldering iron with a clean, tinned tip to remove excess solder balls. Adjust the soldering technique to avoid applying excessive solder.

5. Tombstoning:

Symptoms: One end of a surface-mounted component lifts off the pad, resulting in an uneven joint.

Causes: Imbalanced heat distribution, uneven solder paste application, or mismatched thermal characteristics of the component and the PCB.

Solutions: Ensure even heat distribution during reflow soldering or use a hot air rework station. Apply solder paste evenly, and verify component and PCB thermal compatibility.

6. Solder Joint Cracks:

Symptoms: Fine cracks or fractures in the solder joint.

Causes: Mechanical stress, thermal stress, or poor solder joint quality.

Solutions: Reinforce the joint with additional solder or use strain relief techniques (e.g., using a cable tie to relieve tension). Ensure proper cooling after soldering to minimize thermal stress.

7. Excessive Heat Damage:

Symptoms: Discoloration, burnt PCB, or component damage due to excessive heat.

Causes: Excessive soldering iron temperature, prolonged heating, or inadequate heat dissipation.

Solutions: Use a temperature-controlled soldering iron and set it to an appropriate temperature for the components being soldered. Avoid prolonged heating, use heat sinks or heat-absorbing clips on sensitive components, and ensure proper cooling.

CHAPTER IV. SOLDERING TECHNIQUES

THROUGH-HOLE SOLDERING

Through-hole soldering is a technique used to attach electronic components with leads (wires) to a printed circuit board (PCB) by inserting the leads through holes on the board and soldering them to the corresponding copper pads. This method provides secure mechanical and electrical connections.

Step-by-step guide to through-hole soldering

1. Collect the required tools and equipment:

- Soldering iron with a suitable tip for through-hole soldering.

- Solder wire (appropriate for the components and application).

- Flux (optional, but recommended for better solder flow).

- Desoldering tools (such as desoldering pump or wick) for any potential mistakes or repairs.
- Safety glasses or goggles, heat-resistant gloves, and proper ventilation if needed.

2. Prepare the components and PCB:

- Ensure the PCB and components are clean and free from dirt, oils, or oxidation.
- Trim component leads if necessary and shape them to fit the PCB layout.
- Insert the components into their corresponding holes on the PCB.

3. Heat up the soldering iron:

- Plug in the soldering iron and allow it to heat up to the appropriate temperature for the solder you are using.
- Temperature may vary based on the solder type, so refer to the manufacturer's guidelines.

4. Tin the soldering iron tip:

- Clean the soldering iron tip with a damp sponge or brass wire cleaner.
- Apply a small amount of solder to the tip to create a thin, even coating (tinning).
- This helps with heat transfer and improves soldering efficiency.

5. Apply flux (if necessary):

- Apply a small amount of flux to the joint area to improve solder flow and wetting.
- Flux helps remove oxidation and promotes a strong solder bond.

6. Heat the joint:

- Place the heated soldering iron tip at the junction of the component lead and PCB pad, applying gentle pressure.
- Ensure the tip contacts both the lead and pad simultaneously.
- Heat the joint for a few seconds to allow proper heat transfer.

7. Feed the solder:

- Once the joint is heated, touch the solder wire to the junction of the lead and pad, not the soldering iron tip.
- Let the solder melt and fill the joint.
- Apply enough solder to form a smooth, shiny fillet that covers the entire junction.

8. Remove the solder and iron:

- Remove the solder wire first, followed by the soldering iron.
- Maintain the joint undisturbed until the solder solidifies.
- Avoid moving or touching the joint until it cools down to prevent cold solder joints.

9. Inspect the solder joint:

- Once the joint cools, visually inspect it for a smooth, shiny appearance.
- Check for good wetting and a proper fillet that completely covers the junction.

- Ensure there are no solder bridges or other defects.

10. Repeat for other joints:

- Proceed to solder the remaining through-hole joints following the same steps.
- Take breaks if needed to prevent overheating the components or PCB.

11. Clean and maintain:

- Clean the soldering iron tip with a damp sponge or brass wire cleaner after each joint to remove excess solder.
- Properly store your soldering iron and other tools to ensure their longevity and prevent damage.

Proper solder joint formation

Proper solder joint formation is essential for creating reliable and durable connections in electronic assemblies. Follow these guidelines to achieve high-quality solder joints:

1. Cleanliness: Ensure the surfaces to be soldered are clean and free from dirt, oils, or oxidation. Use isopropyl alcohol and a lint-free cloth to clean the area before soldering.

2. Heat control: Use a soldering iron with a suitable temperature for the solder and components being soldered. High temperatures can damage components or lead to overheating, while low temperatures may result in poor solder flow.

3. Flux application: Apply a small amount of flux to the joint area before soldering. Flux improves wetting by removing oxidation and promotes the flow of solder.

4. Heat transfer: Heat the joint by placing the soldering iron tip at the junction of the component lead and PCB pad. Ensure the tip contacts both the lead and pad simultaneously to provide even heat distribution.

5. Solder application: Touch the solder wire to the junction of the lead and pad, not the soldering iron tip. Allow the solder to melt and flow smoothly into the joint. Use the right amount of solder to create a sufficient fillet that covers the entire junction. Too much solder can cause bridging, while too little solder can result in a weak joint.

6. Solder flow and wetting: Aim for good wetting, which refers to the ability of the molten solder to spread evenly across the joint. It indicates a strong bond and proper contact between the solder and the surfaces being joined. If the solder doesn't flow well, it may indicate an issue with cleanliness, heat, or insufficient flux. Reapply flux, clean the joint, adjust the heat, or ensure proper contact to improve solder flow.

7. Visual inspection: Inspect the solder joint visually after it cools down. A proper solder joint should appear smooth, shiny, and have a distinct fillet that covers the joint fully. Check for any signs of cold solder joints (dull or grainy appearance), solder bridges (excess solder connecting adjacent pads), or insufficient solder (insufficient coverage).

8. Mechanical strength: Ensure the solder joint provides adequate mechanical strength to withstand stress or movement. The joint should securely hold the component in place and resist pulling or twisting forces.

9. Rework and repair: If a solder joint is defective or requires rework, use desoldering tools (such as desoldering pump or wick) to remove the solder and correct the issue. Clean the area and repeat the soldering process as needed.

Desoldering techniques for repairs and rework

Desoldering is the process of removing solder from a joint to correct mistakes, replace components, or perform rework. These are some common desoldering techniques for repairs and rework:

1. Desoldering Pump (Solder Sucker):

- Apply heat to the solder joint using a soldering iron to melt the solder.

- Position the desoldering pump with the plunger fully depressed.
- Place the nozzle of the pump near the molten solder on the joint.
- Release the plunger to create suction and draw the molten solder into the pump.
- Clean the desoldering pump after each use to remove solder residue.

2. Desoldering Braid (Solder Wick):

- Apply heat to the solder joint using a soldering iron to melt the solder.
- Position the desoldering braid over the molten solder on the joint.
- Allow the braid to absorb the molten solder through capillary action.
- Remove the braid once it has absorbed the solder, ensuring it doesn't touch the joint until it cools.
- Trim off the used portion of the braid before each reuse.

3. Hot Air Rework Station:

- Set the hot air rework station to an appropriate temperature.
- Direct the hot air flow onto the solder joint to heat the solder.
- Once the solder is molten, use tweezers or a tool to remove the component or lift the leads from the PCB.
- Clean the remaining solder on the joint using a desoldering pump or braid.

4. Desoldering Iron:

- A desoldering iron combines the functions of a soldering iron and a desoldering pump.
- Heat the solder joint with the desoldering iron to melt the solder.
- Activate the vacuum pump in the desoldering iron to create suction and draw the molten solder into the tool.
- Clean the desoldering iron after each use to remove solder residue.

5. Desoldering Station:

- A desoldering station typically consists of a soldering iron and a built-in vacuum pump for desoldering.
- Use the soldering iron to heat the solder joint until the solder melts.
- Activate the vacuum pump to create suction and remove the molten solder from the joint.
- Clean the desoldering station after each use to remove solder residue.

SURFACE MOUNT SOLDERING

Introduction to surface mount components

Surface mount components, also known as SMD (Surface Mount Device) components, are electronic components that are designed to be directly mounted onto the surface of a printed circuit board (PCB) instead of being inserted through holes like through-hole components. They have become the standard in modern electronics manufacturing due to their compact size, better performance, and automated assembly advantages. Surface mount components are widely used in various electronic devices such as smartphones, tablets, laptops, and many other consumer electronics. Unlike through-hole components, surface mount components have small, flat leads or metal contacts called "pads" that are soldered onto the PCB surface. They are available in various sizes, ranging from tiny components like resistors and capacitors to more complex integrated circuits (ICs) and microcontrollers. The smaller sizes allow for higher component density on the PCB, resulting in more compact and lightweight electronic devices.

Surface mount components offer several advantages:

1. Size and Space Efficiency: SMD components are significantly smaller than their through-hole counterparts, enabling designers to create smaller, more portable devices. This is especially important in modern electronics where miniaturization is a key factor.

2. Enhanced Performance: SMD components often provide better electrical performance due to shorter signal paths and reduced parasitic capacitance and inductance. This can lead to improved speed, efficiency, and overall circuit performance.

3. Automated Assembly: Surface mount technology (SMT) allows for highly automated assembly processes using pick-and-place machines, reflow soldering, and stencil printing. This results in faster production times and lower manufacturing costs compared to manual assembly of through-hole components.

4. Higher Frequencies: SMD components are better suited for high-frequency applications due to their smaller dimensions and reduced parasitics. They can handle higher frequencies without significant signal degradation.

5. Thermal Efficiency: Surface mount components offer improved heat dissipation due to their direct contact with the PCB, which acts as a heat sink. This allows for better thermal management in electronic designs.

6. Cost-Effective: With the advancement of surface mount technology, SMD components have become more cost-effective and widely available, making them a preferred choice in electronics manufacturing.

Soldering surface mount components with different packages

Soldering surface mount components with different packages requires attention to detail and the use of appropriate techniques and tools.

1. Surface Mount Resistors and Capacitors (0402, 0603, 0805, etc.):

- Apply a small amount of solder paste or flux to the pads on the PCB.

- Use tweezers or a vacuum pickup tool to carefully place the component onto the pads, aligning it with the correct orientation.
- Heat the solder paste using a reflow soldering method, such as a hot air rework station or a reflow oven. The solder paste will melt, creating a strong solder joint between the component and the PCB.

2. Small Outline Transistors (SOT-23):

- Apply a small amount of solder paste or flux to the pads on the PCB.
- Use tweezers or a vacuum pickup tool to position the transistor on the pads, aligning it correctly.
- Apply heat to the pads using a soldering iron or hot air rework station. Touch the soldering iron tip to each pad, allowing the solder to melt and create a solid connection. Be careful not to apply excessive heat that could damage the component.

3. Quad Flat No-Lead (QFN) and Dual Flat No-Lead (DFN):

- Apply a small amount of solder paste or flux to the exposed pads on the PCB.
- Use tweezers or a vacuum pickup tool to position the component on the pads, aligning it corrcctly.
- Use a hot air rework station or a reflow oven to heat the PCB. The solder paste will melt, creating a reliable connection between the component and the pads.

4. Ball Grid Array (BGA):

- Apply solder paste to the pads on the PCB using a stencil.
- Carefully place the BGA component on the PCB, aligning it with the correct orientation.
- Use a reflow oven to heat the PCB. The solder paste will melt, creating solder connections between the BGA component and the pads. Ensure that the temperature

profile is suitable for the BGA package to achieve proper solder reflow.

Tips for reflow soldering and hot air rework

Reflow soldering and hot air rework are common methods used for soldering surface mount components.

Reflow Soldering:

1. Temperature Profile: Understand and follow the recommended temperature profile for your solder paste and components. This includes preheating, ramp-up, peak temperature, and cooling stages. Using a reflow oven or a soldering station with temperature control helps achieve precise temperature control.

2. Solder Paste Application: Apply an appropriate amount of solder paste to the pads using a stencil or a syringe. Ensure the paste covers the pads adequately without excessive or insufficient amounts.

3. Component Placement: Carefully position the components on the solder paste-covered pads, ensuring correct orientation and alignment. Using a vacuum pickup tool or tweezers can help with precise placement.

4. Heat Distribution: Ensure even and proper heat distribution during reflow. If using a reflow oven, place the PCB on a tray or fixture designed for proper heat transfer. If reflowing with a hot air rework station, maintain a consistent distance between the nozzle and the PCB to achieve even heating.

5. Thermal Considerations: Take into account the thermal characteristics of the components and the PCB. Avoid rapid temperature changes or thermal shock, as this can damage sensitive components. Preheating the PCB before reflow can help minimize thermal stress.

6. Inspection and Quality Control: After reflow soldering, inspect the solder joints for proper wetting, fillet formation, and overall solder quality. Use a magnifying glass or a microscope to check for any defects, such as bridges, cold joints, or insufficient solder.

Hot Air Rework:

1. Nozzle Selection: Choose the appropriate nozzle size and shape for the component being reworked. The nozzle should fit snugly around the component and evenly distribute hot air.

2. Temperature Control: Set the hot air rework station to the correct temperature suitable for the specific component and solder. Avoid using excessive temperatures that could damage the component or surrounding parts.

3. Airflow and Distance: Adjust the airflow and keep an appropriate distance between the nozzle and the component during rework. Maintain consistent and controlled airflow to evenly heat the component without blowing it off the PCB.

4. Component Removal: Apply heat evenly to the component to melt the solder, and then use tweezers or a vacuum pickup tool to carefully lift and remove the component. Avoid excessive force that could damage the component or the PCB pads.

5. Clean and Inspect: Clean the pads and remove any residual solder or flux using desoldering braid or a desoldering pump. Inspect the pads for any damage or lifted traces. Clean the area before placing a new component.

6. Component Replacement: Apply fresh solder paste or flux to the pads, position the new component with proper alignment, and reflow the solder using the hot air rework station. Ensure the component is correctly seated and soldered to the pads.

CHAPTER V. ADVANCED SOLDERING TECHNIQUES

FINE-PITCH SOLDERING

Challenges and considerations for fine-pitch soldering

Fine-pitch soldering refers to the process of soldering components with very small pin spacing or pitch, typically less than 0.5mm. It presents specific challenges and considerations due to the small size and close proximity of the pins.

1. Component Placement: Accurate component placement is crucial when dealing with fine-pitch components. Use magnification tools such as a microscope or magnifying glass to ensure precise alignment of the component with the PCB pads. Consider using alignment markers or fiducial marks on the PCB to aid in component placement.

2. Solder Paste Application: Apply solder paste with precision to the pads using a stencil or a syringe. Ensure that the paste is applied only to the specific pads to prevent bridging between adjacent pins. Use a thin stencil or a syringe with a fine needle to ensure precise application.

3. Soldering Iron Selection: Use a soldering iron with a fine and pointed tip suitable for fine-pitch soldering. The tip should match the size of the solder pads and pins to ensure proper heat transfer and control during soldering. A temperature-controlled soldering iron is recommended to maintain precise and consistent temperatures.

4. Temperature Control: Fine-pitch components are more sensitive to excessive heat, so accurate temperature control is critical. Use a temperature-controlled soldering station or hot air rework station with fine temperature adjustment capabilities. Avoid prolonged exposure to heat that could damage the component or adjacent parts.

5. Soldering Techniques: Consider using drag soldering or solder reflow techniques for fine-pitch soldering. Drag soldering involves applying a small amount of solder to the tip of the soldering iron and dragging it across the pins, allowing the solder to bridge the gap between the pins and the pads. Solder reflow involves applying solder paste, heating the entire area, and allowing the surface tension of the molten solder to align the pins with the pads.

6. Flux Application: Apply flux to the pads or solder paste before soldering. Flux helps improve solder wetting, reduces oxidation, and enhances solder joint formation. Use a flux pen or a syringe to apply a small amount of flux to the specific areas.

7. Inspection and Quality Control: After soldering, thoroughly inspect the solder joints using magnification tools. Look for proper wetting, fillet formation, and absence of solder bridges or insufficient solder. Conduct electrical testing if necessary to ensure the integrity of the connections.

8. Practice and Patience: Fine-pitch soldering requires practice and patience to develop the necessary skills and hand-eye coordination. Start with simpler components and gradually move on to more complex ones as you gain confidence and experience.

Tools and techniques for precise soldering

When it comes to precise soldering, using the right tools and techniques is crucial. There are some tools and techniques that can help achieve precise soldering:

1. Fine-Tip Soldering Iron: A soldering iron with a fine, pointed tip allows for precise application of heat to specific areas. It provides better control and accuracy when soldering small components or working with tight spaces.

2. Temperature-Controlled Soldering Station: A soldering station with temperature control allows you to set and maintain the appropriate temperature for your soldering task. Precise temperature control is essential for achieving consistent soldering results without overheating or damaging the components.

3. Magnification Tools: Use magnification tools such as a magnifying glass, a magnifying lamp, or a microscope to see the details more clearly. Magnification helps with component placement, solder joint inspection, and working with fine-pitch components.

4. Fine-Tipped Tweezers: Fine-tipped tweezers are handy for holding small components securely during soldering. They provide precise control when positioning components on the PCB and can be used to adjust component alignment during soldering.

5. Helping Hands: Helping hands or a PCB holder with adjustable alligator clips can hold the PCB in place, keeping it steady while you solder. This provides stability and frees up your hands to focus on precise soldering.

6. Soldering Flux: Flux is essential for improving solder wetting and enhancing solder joint formation. Use a flux pen or a syringe to apply a small amount of flux to the soldering area. Flux helps to remove oxidation, improve solder flow, and ensure reliable connections.

7. Solder Wick or Desoldering Pump:
Desoldering tools like solder wick (soldering braid) or a desoldering pump are useful for removing excess solder or correcting soldering mistakes. They help maintain clean and precise solder joints.

8. Fine-Gauge Solder: Using a fine-gauge solder wire, such as 0.015" or 0.020", enables precise application of solder to small pads and pins. The thinner solder wire allows for better control and prevents excessive solder buildup.

9. Practice and Patience: Precise soldering skills are developed through practice and patience. Start with simpler soldering tasks and gradually progress to more complex ones as you gain confidence and refine your technique.

10. Clean and Tidy Workspace: Maintain a clean and tidy workspace to avoid distractions and ensure better concentration during soldering. A clutter-free area allows you to focus on the task at hand and minimizes the risk of damaging components or causing soldering errors.

WIRE AND CABLE SOLDERING

Soldering wires and cables for connections and repairs

Soldering wires and cables is a common practice for making reliable electrical connections and repairs.

1. Prepare the wires: Strip about 1/2 to 3/4 inch (1.3 to 1.9 cm) of insulation from the ends of the wires you want to solder. Use a wire stripper or a sharp knife, being careful not to cut into the wire strands.

2. Twist the wire strands: For stranded wires, gently twist the individual strands together to ensure they are secure and prevent fraying. This step is not necessary for solid core wires.

3. Apply soldering flux: Apply a small amount of soldering flux to the exposed wire strands. Flux helps improve solder wetting and ensures a reliable solder joint.

4. Pre-tin the wires: Heat your soldering iron and touch it to the wire strands. Feed a small amount of solder onto the wire and allow it to flow and coat the strands evenly. This process is known as pre-tinning and helps facilitate the soldering process.

5. Prepare the cable (if applicable): If you are soldering a cable, separate the individual wires and strip a small amount of insulation from each wire. Twist the strands of each wire separately.

6. Align and join the wires: Align the pre-tinned wire strands and cables (if applicable) so that they overlap by about 1/4 to 1/2 inch (0.6 to 1.3 cm). Make sure the wires are properly aligned for a secure connection.

7. Heat the wires: Hold the soldering iron tip against the wire junction, applying heat evenly to the wires. Be careful not to overheat or damage the insulation around the wires.

8. Apply solder: Once the wires are heated, touch the solder wire to the junction of the wires, allowing the solder to melt and flow around the wires. Ensure that the solder flows into the joint and coats the wire strands evenly.

9. Remove the soldering iron: Remove the solder wire and then remove the soldering iron tip from the joint, but hold the wires steady until the solder cools and solidifies.

10. Inspect and protect: Inspect the solder joint to ensure it is smooth, shiny, and free from any cold solder joints, bridges, or excessive solder. If necessary, use desoldering tools to correct any mistakes. Once the joint is solid and cooled, you can use heat shrink tubing, electrical tape, or other appropriate insulation methods to protect the solder joint and ensure electrical safety.

Splicing and insulation techniques

When splicing wires, it's important to ensure a secure and insulated connection. There are some techniques for splicing wires and insulating the connection:

1. Overlap and Twist: Align the stripped ends of the wires you want to splice. For solid core wires, overlap the ends by about 1 inch (2.5 cm). For stranded wires, overlap them by about 1/2 to 3/4 inch (1.3 to 1.9 cm). Twist the wire strands together in a clockwise direction using pliers or your fingers. Ensure a tight and secure twist.

2. Inline Sleeve: Slide a heat shrink sleeve over one of the wires before twisting the splice. Make sure the sleeve is long enough to cover the entire splice area. After twisting the wires, position the sleeve over the splice area, covering it completely.

3. Soldering: Apply soldering flux to the twisted wire strands to enhance solder flow and wetting. Heat the wires with a soldering iron until they reach the solder's melting point. Apply solder to the heated area, allowing it to flow into the splice and coat the wire strands evenly. Remove the soldering iron and let the joint cool.

4. Insulation Tubing: Slide a larger diameter heat shrink tubing over the entire splice area, ensuring it covers the solder joint and extends beyond the wire insulation. Heat the tubing with a heat gun or a heat source until it shrinks tightly around the splice, providing insulation and strain relief.

5. Electrical Tape: Alternatively, you can use electrical tape to insulate the splice. Begin by wrapping a layer of electrical tape tightly around the splice area, ensuring it covers the entire joint and extends beyond the wire insulation. Wrap additional layers of tape to build up insulation and provide mechanical protection.

6. Liquid Electrical Tape: Another option is to use liquid electrical tape. Apply a coating of liquid electrical tape over the splice, ensuring it covers the joint and extends beyond the wire insulation.

Allow the liquid electrical tape to dry and form a flexible, protective coating.

7. Wire Nuts or Crimp Connectors: Wire nuts or crimp connectors can be used for splicing wires in certain applications. Twist the stripped wire ends together and secure them using a wire nut or crimp connector. Ensure a tight connection and follow the manufacturer's instructions for proper installation and insulation.

SOLDERING CIRCUIT BOARDS

Circuit board preparation and cleaning

Proper preparation and cleaning of circuit boards are essential for achieving reliable solder joints and ensuring the overall functionality of electronic assemblies. These are some steps to follow for circuit board preparation and cleaning:

1. Inspect the Circuit Board: Before starting any preparation or cleaning, inspect the circuit board for any visible defects, such as damaged traces, lifted pads, or components that need rework.

Address these issues before proceeding with the preparation and cleaning process.

2. Gather Necessary Materials: Gather the required materials for circuit board preparation and cleaning, which may include:

- Isopropyl alcohol (IPA): Use high-purity IPA (preferably 99% or higher) for cleaning purposes.
- Lint-free wipes or swabs: Choose lint-free wipes or swabs that won't leave residues or fibers on the board.
- Soft-bristle brush: Use a soft-bristle brush to remove dust and debris from the board's surface and components.
- Compressed air: Use compressed air to blow away loose debris and particles from the board.

3. Remove Dust and Debris: Start by gently brushing the circuit board's surface and components with a soft-bristle brush to remove any loose dust or debris. Be careful not to apply excessive force that could damage the board or components. Alternatively, you can use compressed air to blow away the debris.

4. Clean the Circuit Board: Moisten a lint-free wipe or swab with isopropyl alcohol (IPA) and gently wipe the board's surface and components. IPA is effective in removing contaminants such as grease, oils, flux residues, and other organic substances. Pay close attention to areas where flux residues or other contaminants are present, ensuring thorough cleaning.

5. Rinse (if necessary): If the circuit board has been exposed to water-soluble flux during soldering, it may require rinsing. In such cases, use deionized water to rinse the board and remove any water-soluble flux residues. Ensure the board is completely dry before proceeding.

6. Inspect the Cleaned Board: After cleaning, inspect the circuit board again to ensure it is free from visible contaminants, residues, or debris.

Check for any areas that may require additional cleaning or rework.

7. ESD Precautions: Throughout the preparation and cleaning process, follow appropriate electrostatic discharge (ESD) precautions to prevent static damage to sensitive components or the circuit board itself. Use grounded ESD mats, wrist straps, and other ESD-safe equipment.

Soldering multi-pin components and ICs

Soldering multi-pin components and integrated circuits (ICs) can be more challenging due to the higher number of pins and their close proximity.

1. Gather the necessary tools and materials: Prepare the appropriate soldering iron, solder, flux, solder wick or desoldering pump (in case of mistakes), and any necessary aids such as a PCB holder or helping hands.

2. Prepare the PCB: Ensure the PCB is clean and free from debris or contaminants. If necessary, clean the PCB using isopropyl alcohol and a lint-free cloth.

3. Component Placement: Carefully position the multi-pin component or IC on the PCB, aligning it with the corresponding pads or holes. Double-check the orientation and alignment to avoid errors.

4. Tack Soldering: Apply a small amount of solder to one corner pad or pin of the component. This tack soldering helps hold the component in place while you solder the remaining pins. Ensure the component is securely held during this step.

5. Soldering the Remaining Pins: Starting from the tack-soldered corner, work your way across the component, soldering each pin one at a time. Heat the pad and pin simultaneously with the soldering iron for a few seconds until the solder flows and forms a smooth, shiny solder joint. Be careful not to overheat the component or adjacent pins.

6. Soldering Technique: Use a technique such as drag soldering or solder bridging to solder multiple pins quickly and efficiently. For drag soldering, apply flux to a row of pins, touch the soldering iron tip to one end of the row, and then drag the solder wire along the pins while the solder melts and flows into the joints. Remove excess solder using solder wick or a desoldering pump if needed.

7. Inspect the Solder Joints: Once all the pins are soldered, visually inspect the solder joints. Ensure they are smooth, shiny, and free from solder bridges, cold joints, or excessive solder. Use a magnifying glass or microscope if necessary for closer inspection.

8. Clean and Remove Flux Residue: Clean the soldered joints and PCB to remove any flux residue. Flux residues can cause corrosion and affect the electrical performance of the circuit. Use isopropyl alcohol and a brush or lint-free cloth to clean the area. Ensure the board is completely dry before further testing or handling.

Soldering wires and jumpers on circuit boards
Soldering wires and jumpers on circuit boards is a common practice for making connections or adding additional wiring to a circuit. Here are the

steps to solder wires and jumpers onto a circuit board:

1. Prepare the wires or jumpers: Cut the wires or jumpers to the desired length, leaving some extra length for flexibility and ease of connection. Strip about 1/4 to 1/2 inch (0.6 to 1.3 cm) of insulation from the ends of the wires or jumpers using a wire stripper or a sharp knife.

2. Prepare the circuit board: Ensure the circuit board is clean and free from debris or contaminants. If necessary, clean the board using isopropyl alcohol and a lint-free cloth.

3. Identify the connection points: Identify the specific locations on the circuit board where you want to solder the wires or jumpers. This can be pads, through-holes, or specific solder points.

4. Tin the wires: Apply soldering flux to the exposed wire strands. Heat the wire with a soldering iron and apply a small amount of solder to the wire strands, allowing it to melt and flow evenly. This process is known as tinning and helps improve solder wetting and ensure a reliable solder joint.

5. Preheat the connection points: Preheat the connection points on the circuit board using a soldering iron. This will help the solder flow and adhere properly when making the connections.

6. Solder the wires or jumpers: Place the tinned wire or jumper onto the preheated connection point, ensuring a good alignment and contact. Apply the soldering iron tip to the wire or jumper and the connection point simultaneously. Feed a small amount of solder onto the joint and allow it to melt and flow, creating a solid solder joint. Ensure the solder flows into the joint and coats the wire or jumper and the connection point evenly.

7. Inspect and clean: Once the solder joints are made, visually inspect them to ensure they are smooth, shiny, and free from any cold solder joints, bridges, or excessive solder. Use a magnifying glass or microscope if necessary for closer inspection. Clean the soldered area and the circuit board using isopropyl alcohol and a brush or lint-free cloth to remove any flux residue.

8. Test the connections: After soldering, test the connections using appropriate electrical testing methods to ensure proper conductivity and functionality.

CHAPTER VI. TROUBLESHOOTING AND REPAIR

IDENTIFYING AND FIXING COMMON SOLDERING DEFECTS

Soldering defects can occur during the soldering process, and it's important to identify and fix them to ensure reliable and effective connections. There are some common soldering defects, along with their identification and potential solutions:

1. Cold Solder Joint:

Identification: A cold solder joint appears dull, grainy, or lumpy. It may have incomplete wetting, with solder not flowing properly onto the joint, resulting in a weak connection.

Solution: To fix a cold solder joint, reheat the joint using a soldering iron and apply fresh solder. Ensure the joint is heated adequately to allow the solder to flow and wet the joint properly.

2. Solder Bridging:

Identification: Solder bridging occurs when excess solder forms an unintended connection between two adjacent pins or pads, resulting in a short circuit.

Solution: To fix solder bridging, use desoldering wick or a desoldering pump to remove excess solder. Ensure the soldering iron is clean and properly tinned before re-soldering the joints, keeping thc solder confined to each individual connection.

3. Insufficient Solder:

Identification: Insufficient solder results in a thin, weak joint with poor electrical conductivity. The joint may appear dull or have exposed wires or pins.

Solution: To fix insufficient solder, reheat the joint with a soldering iron and add a small amount of additional solder. Ensure the solder flows and covers the joint adequately, creating a robust connection.

4. Excessive Solder:

Identification: Excessive solder forms a bulky or blob-like joint that may extend beyond the intended connection area. It can cause short circuits or interfere with adjacent components.

Solution: To fix excessive solder, use desoldering wick or a desoldering pump to remove the excess solder carefully. Clean the area, re-tin the joint, and apply the appropriate amount of solder, ensuring it is within the designated connection area.

5. Solder Ball:

Identification: Solder balls are small blobs of solder that form separate from the joint, often caused by excessive solder or improper soldering technique.

Solution: To fix solder balls, use a soldering iron with a fine tip to reheat the joint. Draw the soldering iron away from the solder ball, causing it to reflow and merge back into the joint. Use desoldering wick or a desoldering pump if necessary to remove excess solder.

6. Lifted Pad:

Identification: A lifted pad occurs when the pad

on the circuit board detaches or lifts from the surface, making it difficult to create a reliable solder connection.

Solution: To fix a lifted pad, gently scrape away any damaged or loose traces around the pad. Clean the area and use wire or copper tape to create a bridge between the lifted pad and a nearby trace or component lead. Solder the wire or copper tape to establish a new connection.

REPAIRING DAMAGED SOLDER JOINTS AND CONNECTIONS

Repairing damaged solder joints and connections is necessary to restore functionality to electronic components or circuits.

1. Identify the damaged solder joint or connection: Inspect the component or circuit to identify the specific solder joint or connection that needs repair. Look for signs of physical damage, such as a cracked joint, loose wire, or detached component.

2. Prepare the area: Clean the damaged area using isopropyl alcohol and a lint-free cloth or brush to remove any debris, flux residues, or oxidation. Ensure the area is dry before proceeding.

3. Remove the damaged solder: If necessary, use a desoldering pump or desoldering wick to remove the old solder from the damaged joint. Heat the joint with a soldering iron while applying the desoldering tool to suck out the molten solder or use the desoldering wick to absorb it.

4. Clean the area: Clean the area again using isopropyl alcohol to remove any remaining flux residues or contaminants. Ensure the area is dry before continuing.

5. Prepare the solder joint: Apply flux to the cleaned area to promote proper solder wetting and improve the soldering process. Flux helps remove oxidation and ensures a reliable solder connection.

6. Position the component or wire: Place the component or wire in its correct position, aligning it with the corresponding pads or connection points. Make sure it is securely positioned and held in place, either by hand or with the help of a PCB holder or helping hands.

7. Solder the joint: Heat the solder joint with a soldering iron and apply a small amount of solder to the joint. Allow the solder to flow and create a smooth, shiny solder joint. Ensure the solder fully wets the component lead, wire, or pad, providing a strong connection.

8. Inspect the repaired joint: After soldering, visually inspect the repaired joint to ensure it is properly formed, smooth, and free from defects or excess solder. Use a magnifying glass or microscope if necessary for a closer examination.

9. Test the repair: After completing the repair, test the component or circuit to ensure the repaired solder joint or connection is functioning correctly. Check for continuity, proper electrical connections, or any other specific functionality based on the component or circuit being repaired.

TROUBLESHOOTING FAULTY SOLDERED CIRCUITS

Troubleshooting faulty soldered circuits can help identify and fix issues that may arise after soldering.

1. Visual Inspection: Begin by visually inspecting the soldered joints and connections. Look for any obvious defects such as cold joints, solder bridges, or insufficient solder. Check for misaligned components, damaged traces, or loose wires.

2. Check for Continuity: Use a multimeter to check for continuity between the appropriate points in the circuit. Ensure that the soldered connections provide a proper electrical path and that there are no open circuits or short circuits.

3. Check for Short Circuits: Use a multimeter set to the continuity or resistance mode to check for unintended short circuits between adjacent soldered connections or traces. Look for solder bridges or unintentional connections that may cause electrical issues.

4. Reheat and Resolder: If you suspect a particular solder joint to be faulty, reheat and resolder it. Apply flux, heat the joint with a soldering iron, and add fresh solder to ensure a proper connection.

5. Inspect Component Orientation: Verify that all components are correctly oriented and placed on the board. Incorrect component orientation can lead to circuit malfunctions.

6. Test Components: Test individual components using appropriate testing equipment or substitute components if necessary. Faulty components can sometimes cause circuit issues that may be mistakenly attributed to soldering.

7. Check for Mechanical Stress: Ensure that there is no excessive mechanical stress on soldered connections or components. Stress can lead to cracked joints or intermittent connections.

Reinforce any weak or vulnerable areas with additional solder or suitable mechanical support.

8. Verify Power Supply: Check the power supply voltage and current to ensure it meets the required specifications for the circuit. Inadequate power supply can cause circuit malfunctions.

9. Consult Schematics or Documentation: Refer to the circuit schematics or documentation to verify the correct connections and functionality of the circuit. Ensure that the soldered connections align with the intended design.

10. Reassemble and Test: Once you have addressed any identified issues, reassemble the circuit and perform thorough testing to verify its functionality. Test various functions and inputs to ensure proper operation.

www.ingramcontent.com/pod-product-compliance
Lightning Source LLC
Chambersburg PA
CBHW072032230526
45466CB00020B/1738